Topical Conference: Sustainable Pathways Toward Hydrogen and Synthetic Fuels

Held at the 2023 AIChE Annual Meeting

Orlando, Florida, USA
5-10 November 2023

ISBN: 978-1-7138-9314-1

Printed from e-media with permission by:

Curran Associates, Inc.
57 Morehouse Lane
Red Hook, NY 12571

AIChE
120 Wall Street, FL 23
New York, NY 10005-4020

Phone: (800) 242-4363
Fax: (203) 775-5177

www.aiche.org

Additional copies of this publication are available from:

Curran Associates, Inc.
57 Morehouse Lane
Red Hook, NY 12571 USA
Phone: 845-758-0400
Fax: 845-758-2633
Email: curran@proceedings.com
Web: www.proceedings.com

TABLE OF CONTENTS

SUSTAINABLE PATHWAYS TO CLEAN HYDROGEN AND SYNTHETIC FUELS IV

SUSTAINABLE PATHWAYS TO CLEAN HYDROGEN AND SYNTHETIC FUELS V

12a US Department of Energy Hydrogen Program Overview

Eric Miller, William Gibbons

Today, technologies for advancing National clean energy priorities are rapidly evolving, including hydrogen and fuel cell technologies which offer unique versatility within a portfolio of domestic options addressing decarbonization, economic growth, and environmental justice. The U.S. Department of Energy's Hydrogen Program, in support of the Hydrogen Energy Earthshot and the H2@Scale initiative, comprises a broad portfolio of research, development, demonstration, and deployment (RDD&D) focused on technologies for the affordable production, storage, distribution, and utilization of clean hydrogen across sectors. Coordinated by the Department's Hydrogen and Fuel Cell Technologies Office, this portfolio includes RDD&D focused on decarbonizing industrial and chemical end uses, including the sustainable production of clean and affordable fuels and byproducts. This talk provides an overview of the DOE Hydrogen Program and its alignment with National energy and environmental priorities; and highlights specific recent accomplishments in the area of sustainable fuels.

12b H2NEW Lab Consortium: Increasing the Performance and Durability of Electrolysis for Clean, Competitive Production of Hydrogen

Richard Boardman, Bryan S. Pivovar, David Peterson, William Gibbons, Mark Ruth, Daniel Wendt, Olga Marina, Rangachary Mukundan

The H2NEW (Hydrogen from Next-generation Electrolyzers of Water) consortium aims to overcome technical barriers to affordable, reliable, and efficient electrolyzers of water—empowering industry with insights for achieving low-cost and large-scale clean hydrogen production. The primary focus of H2NEW is to understand and prove the long-term durability of electrolysis at target performance conditions (power to hydrogen production efficiency), thus, to enable hydrogen production from electrolysis for use as a clean, sustainable fuel or feedstock for chemicals and combustible fuels production. The overarching goal is to help industry produce hydrogen with clean energy inputs for less than \$2/kg-H2 by 2026, and then reaching \$1/kg-H2 by 2031. At these price points, hydrogen can be combined with a broad spectrum of recycled carbon sources to produce sustainable fuels, beginning with fungible liquid hydrogen carbon fuels, and progressing to substitute natural gas (SNG). With the buildup of zero-emissions power generation, electrolysis could legitimately replace a sizeable fraction of the conventional fuels used by industry and the transportation sectors.

H2NEW is advancing a deeper science-based understanding of liquid-alkaline (L-AE), proton-exchange membrane (PEM), and oxygen-ion conduction solid-oxide cells (O-SOEC) through integrated materials testing, characterization, and modeling and simulation. Accelerated stress testing methods and are being applied to gain an understanding of microstructure evolution. Multi-scale modeling is used to inform cell testing conditions and to gain a deeper understanding of the materials behavior as a function of cell designs and operating conditions. This work is helping industry and academic collaborators understand and mitigate cell degradation while pushing performance higher with this understanding.

12c Doe Thermal Conversion with Carbon Capture and Storage RandD

Eva Rodezno

This is an unprecedented time in history for hydrogen with interest being amplified worldwide due to its potential to address the climate crisis as well as energy security and resiliency. Though there are major challenges, zero and low-carbon hydrogen can be a key part of a comprehensive portfolio of solutions to achieve a sustainable and equitable clean energy future. The US Department of Energy's Fossil Energy and Carbon Management Office is investing significant R&D in low-carbon hydrogen derived from fossil, waste and biomass feedstocks, primarily through gasification, coupled with carbon capture and storage, and clean hydrogen from solid oxide electrolyzers. The Office's efforts are an integral part of the Department of Energy's (DOE) recently launched Hydrogen Shot, with a goal of reducing clean hydrogen costs by 80% to $1 per 1 kilogram (kg) within 1 decade (1-1-1). For gasification systems, DOE is investing in: **Using novel microwaves for enhanced gasification robustness and quality**: This activity area aims to advance waste cleanup/gasification technology by developing microwave enhancements that produce valuable hydrogen from organic impurities in syngas that might survive the primary gasification process. Addressing the concerns of condensable organic species in raw syngas increases the likelihood of commercial sector adoption of flexible and variable waste-feedstocks to clean hydrogen production. Incorporating microwave technology at a gasifier's exit offers the promise of converting any remaining complex molecules into simple molecules like hydrogen, which will improve the systems reliability, availability, and maintainability, all of which allows greater impact on generating clean electricity with net-zero carbon emissions.

Enhanced biomass pretreatment processes: This activity area aims to develop, investigate, and advance the TRL of various novel biomass pretreatment techniques. Biomass-based systems are severely limited in scope due to two key issues: feeding materials into the gasification process, and accessing, transporting, and storing the material in a low- or zero-carbon method, from a lifecycle perspective. Development of these technologies focuses on addressing these two key issues.

Clean hydrogen production systems integration & optimization: This activity area focuses on integration of components within systems, hybridization of processes, and improving process reliability, to optimize hydrogen production efficiency and cost reductions given scale of the system, feedstock availability, and market factors.

Scale-up of moderate-TRL advanced oxygen production technologies: The activity area aims to advance the Technology Readiness Level (TRL) of emerging oxygen production technologies from air separation. Scale-up will allow for maturation of novel, highly efficient, and lower cost oxygen production technologies from DOE's national labs into commercially-relevant prominence, thus facilitating interest from the commercial sector and attracting technology transfer partner(s). Lower cost oxygen is crucial to enable gasification to leverage pre-combustion carbon capture and produce clean hydrogen at a low enough cost to meet the Hydrogen Shot initiative's goal. This technological maturation activity will improve the viability and economics of net-zero carbon gasification systems.

An additional important aspect of low-carbon hydrogen production from these sources is the carbon capture component. Key RDD&D challenges in this space include: Improving Scalability – providing economic viability at all relevant process scales; Optimizing Thermodynamics – reducing energetic requirements through better regeneration energy, lower pressure drops, lower required temperatures, and process optimization; Improving Kinetics – improving equipment through faster, more selective chemical/physical separation pathways; Reducing Capital Cost – reducing equipment size and costs through advanced manufacturing, process intensification, integration, and optimization; Improving Durability – providing rugged long-term performance with slow degradation rates; Improving Flexibility – improving process dynamics by improving turn down and operation at variable capture rates; Minimizing Environmental Impact – providing technologies that minimize air pollution release and minimizing waste generation. DOE's Office of FECM will be funding many Front End Engineering Design projects and pilot scale CCS projects. R&D for CO2 capture technologies such as non-aqueous solvents, membranes, advanced sorbents, and cryogenic processes continues and may lead to significantly decreased energy needs. DOE will rely on advanced computational tools for rational material discovery, design of advanced capture systems components, use of advanced manufacturing, and synthesis of these materials with characterization of their physical properties.

Besides the work being funded by FECM on clean hydrogen production, FECM is also funding CO2 conversion to fuel technologies. Each conversion technology comes with challenges and opportunities. A critical challenge across conversion technology pathways (catalytic conversion and bio-mediated) is the cost-effective, energy-efficient, and selective upgrading of CO2. CO2 is a stable, non-reactive molecule that typically requires heat or electricity, and other reactants to be converted into products. This program works to address the need for enabling technologies, including using carbon-neutral hydrogen as a reactant in the synthesis of fuels and chemicals. The efficiency of reaction conversion and energy use of these utilization processes also represent critical challenges. The Carbon Conversion program supports lab- and bench-scale carbon conversion technologies that have the potential to develop carbon-based products that promise GHG and environmental benefits over incumbent products. Areas of research include, but are not limited to, new projects focused on the catalytic conversion to higher value products such as fuels, chemicals, and polymers; generation of synthetic aggregates; and algal systems with high CO2 utilization efficiency of conversion to various bioproducts. The program aims to continue investment activities from the past, such as reactive capture and conversion, and progress first generation conversion technologies to field-scale testing. Additional efforts will include guidance on benchmarking prototypical catalytical conversion, such as electrochemical reduction for carbon conversion, as well as developing techno-economic analysis (TEA) guidance for screening various technology pathways or product markets.

DOE FECM's is also investing in Reversible Solid Oxide Fuel Cells (R-SOFCs) which can use 100% hydrogen or hydrocarbon fuels to produce electricity, water and CO_2 when operating in a fuel cell mode. R-SOFCs can be configured to operate in reverse as an electrolyzer using power and water as inputs to produce hydrogen, with oxygen as a byproduct. This electrolyzer mode turns the R-SOFC into a Solid Oxide Electrolyzer Cell (SOEC). SOECs essentially function as an SOFC in reverse and optimize the use of these system to reduce overall costs. The carbon dioxide produced from the process with natural gas as a fuel in a fuel cell mode can then be sequestered for storage or use in other applications.

R-SOFCs can both store and produce energy with a single system and can contribute to clean energy generation/storage when paired with a renewable fuel such as hydrogen (in fuel cell mode) or renewable electricity (in electrolysis mode). Hydrogen created from R-SOFCs is a promising fuel source and can be stored for future use when renewable energy sources are not available. When the grid demands power, the R-SOFC consumes the stored hydrogen to produce electricity. R-SOFCs allow for a continuous stream of clean energy into the grid.

The focus areas for Reversible Solid Oxide Fuel Cells Program include:

- Developing and validating the materials proposed for improving the cost, performance, and reliability of R-SOFC systems; and
  RDD&D for degradation at start-up of SOEC operation and enabling technologies for dynamic

- operation of SOEC/SOFC Systems.

12d U.S. Doe Overview - Supporting Clean Hydrogen through Supply Chain and Recycling Research and Development Efforts

Julie Fornaciari

Today, technologies for advancing National clean energy priorities are rapidly evolving, including hydrogen and fuel cell technologies which offer unique versatility within a portfolio of domestic options addressing decarbonization, economic growth, and environmental justice. Through the Bipartisan Infrastructure Law, there is a major focus on developing a robust, domestic supply chain for clean hydrogen technologies through research and development. Additionally, efforts to find ways to sustainably recover and recycle components of the clean hydrogen technologies will reduce the footprint and the impact of resources. Clean hydrogen technologies include heavy duty fuel cells and water electrolyzers as well as industrial and chemical end uses, like steel and ammonia. This talk will provide an overview of the Bipartisan Infrastructure Law program within the U.S. Department of Energy's Hydrogen and Fuel Cell Technologies Offices including the how the work fits into the overall strategy for promoting clean hydrogen as a sustainable fuel.

12e Renewable to Liquid Fuels Beyond the Grid

James Seaba

In order to meet current climate goals, implementation of renewables must double every 5 years. The construction of renewables (wind, solar), however, is hampered by challenges to connect to the electric grid. Wait times for wind and solar projects now extend out to four years. The combined solar projects waiting in line could power an estimated 5.6 million homes. How can we accelerate the building of new solar and wind projects without the use of the grid?

The reduction of carbon emissions by 2050 to net zero is a very challenging task. For example, the transportation sector represents 27% of our current GHG emissions. The majority of this sector can utilize batteries for transportation that represents 62% of transportation. This assumes a significant build-out of the grid system using clean energy to support this massive energy need, that will further exacerbate the long wait times. The difficult to decarbonize transportation sector, mainly aviation and heavy duty vehicles are researching alternative energy sources with no clear economic winner. Synthetic fuels are an obvious option, but very expensive. How can we economically generate 100% sustainable synthetic fuels that address the difficult to decarbonize end-use devices like aviation and semi-trucks?

The answer to both of these questions may be using renewables to make liquid fuels and fuel feedstocks. The economic driver for this process is cheap clean hydrogen (<$2/kg). The carbon is provided by bio-genic processes, ex. Ethanol production, or direct air capture. The cost of the carbon dioxide needs to be approximately $100/ton to help meet cost targets. The power to synthetic liquids using renewables beyond the grid enables cheap electrons, with the final product within market value norms. However, the intermittent source of power coupled with small scale production (100 to 1000 BBL/D) are challenging program metrics.

This paper lays out the vision of a renewable to liquids program and the technology options that may enable this program to be a success. Significant challenges will be highlighted.

115a Dynamic Optimization of Proton Exchange Membrane Water Electrolzyers Considering Usage-Based Degradation

Landon Schofield, Benjamin Paren, Ruaridh Macdonald, Yang Shao-Horn, Dharik Mallapragada

Electrolytically produced hydrogen (H_2) continues to gain traction as a means of producing low-carbon hydrogen. As compared to other low temperature electrolyzers, proton exchange membrane (PEM) systems are particularly promising because of their small size, ability to operate at higher current densities (≥ 1 A cm^{-2}), and ability to operate with a differential pressure between the anode and cathode, which enables production of high pressure H_2 product. Several techno-economic optimization studies have highlighted the importance of dynamic operation of PEM electrolyzers to minimize the levelized cost of hydrogen when utilizing electricity sourced from variable renewable energy (VRE) electricity sources that are co-located or supplied via connection to the electric grid. These studies reveal that cost-optimal dynamic operation involves operating operating at high current densities ($\gg 2$ A cm^{-2}) during time of abundant VRE supply or low electricity prices, as well as periods of idling or near-zero current densities (during high electricity prices or low VRE supply periods), that collectively results in smaller stack areas when compared to an equivalent steady state electrolyzer operating at current densities near 1-2 A cm^{-2}.

Though more efficient for capital and operating costs, dynamic operation tends to lead to faster stack degradation and shortened lifetime in comparison to electrolyzers operating at steady state [1]. Operation at high current densities leads to higher rates of stack degradation, while PEM electrolyzers left to operate at near open circuit conditions have shown accelerated degradation as well [1]. Most techno-economic studies overlook such use-dependent degradation that would inevitably result from dynamic stack operation [2]. Here, we develop a first-principles, 0-D electrochemical model for PEM electrolysis to simulate stack level operation over time while tracking key process variables, such as temperature fluctuations and species concentrations at the anode/cathode. To characterize use-dependent stack degradation in our modeling, we develop an empirical relation to quantify marginal degradation rates as function of current density that is able to approximate experimental trends observed in the literature related to accelerated degradation at low and high current densities [1, 3–8].

We embed the above-mentioned first-principle stack model and degradation correlation into an techno-economic optimization model to evaluate least-cost system design and operation under time-varying electricity prices. The cost-optimization model, a nonlinear program (NLP) implemented in Pyomo, considers the following important operating constraints: a) hourly production requirements for H_2, met through a combination of electrolyzer operation and on-site H_2 storage discharge that was previously charged from the electrolyzer output, b) operating temperature limits for the stack, and c) ensuring H_2 concentration in the anode is below 2 vol% for safe operation [9]. The model is evaluated for various scenarios of future electricity price conditions as well as electrolyzer capital costs and stack replacement costs, to understand the value of dynamic operation vs. steady-state operation considering stack degradation cost impacts.

The presentation will cover a summary of the analysis with a particular emphasis on characterizing the type and range of dynamic operation that would be most economically viable relative to static operation when considering degradation associated trade-offs and electricity price volatility.

References

1. Na Li, Samuel Simon Araya, and Søren Knudsen Kær. "Investigating low and high load cycling tests as accelerated stress tests for proton exchange membrane water electrolysis". In: Electrochim. Acta 370 (Feb. 2021), p. 137748.

2. Brian D James, Daniel A DeSantis, Genevieve Saur. Final Report: Hydrogen Production Pathways Cost Analysis (2013 – 2016). Tech. rep. Strategic Analysis Inc, Sept. 2016.

3. Michel Suermann, Boris Bensmann, and Richard Hanke-Rauschenbach. "Degradation of proton exchange membrane (PEM) water electrolysis cells: Looking beyond the cell voltage increase". en. In: J. Electrochem. Soc. 166.10 (2019), F645–F652.

4. Christoph Rakousky et al. "Polymer electrolyte membrane water electrolysis: Restraining degradation in the presence of fluctuating power". en. In: J. Power Sources 342 (Feb. 2017), pp. 38–47.

5. Georgios Papakonstantinou et al. "Degradation study of a proton exchange membrane water electrolyzer under dynamic operation conditions". en. In: Appl. Energy 280.115911 (Dec. 2020), p. 115911.

6. Shaun M Alia, Sarah Stariha, and Rod L Borup. "Electrolyzer Durability at Low Catalyst Loading and with Dynamic Operation". en. In: J. Electrochem. Soc. 166.15 (Oct. 2019), F1164.

7. Shucheng Sun et al. "Investigations on degradation of the long-term proton exchange membrane water electrolysis stack". In: J. Power Sources 267 (Dec. 2014), pp. 515–520.

8. Steffen Henrik Frensch et al. "Influence of the operation mode on PEM water electrolysis degradation". In: Int. J. Hydrogen Energy 44.57 (Nov. 2019), pp. 29889–29898.

9. Bethany Nicholson et al. "pyomo.dae: a modeling and automatic discretization framework for optimization with differential and algebraic equations". In: Math. Program. Comput. (2018).

115b Modeling the Effect of Material Properties on Liquid Alkaline Water Electrolyzer Performance

Eric Lees, Alexis T. Bell, Adam Weber

Liquid alkaline water electrolyzers (LAWEs) have been used to produce H_2 in industry using excess electricity since the early 1900's. These devices benefit from highly-stable and inexpensive porous nickel electrodes that are used to mediate the anodic oxygen evolution reaction (OER) and cathodic hydrogen evolution reaction (HER). However, LAWEs rarely achieve high reaction rates (i.e., current densities) because of high Ohmic resistances in the porous separator, slow reaction kinetics, and a reduction in nickel catalyst surface area caused by H_2 and O_2 bubble formation during operation. New porous electrodes, separators, and operational philosophies must be developed to manage these issues and increase the current density of LAWEs.

This talk will discuss a 1D continuum model of a LAWE that combines the Nernst-Planck equation with Butler-Volmer kinetics to resolve pH and conductivity gradients in LAWEs. The simulation deconvolutes experimentally-measured applied voltages into kinetic, Ohmic, and concentration polarization losses. The model was applied to several data sets to explore the effect of nickel foam properties (i.e., specific surface area and porosity) and separator resistance on the current density of LAWEs. The model shows that the pore diameter and porosity of nickel foams critically impacts the electrochemically-active surface area through both geometric and bubble coverage effects. The separator was found to contribute most significantly to the total Ohmic resistance, but this resistance can be reduced by decreasing the separator tortuosity and thickness. This talk will provide an overview of these structure-property-performance relationships that can be used to inform the design of advanced electrodes and separators for LAWEs.

115d Engineering the Water Balance in Hydroxide Exchange Membrane Electrolyzers Under Cathode Dry Conditions

Alexandra Oliveira, Brian P. Setzler, Fei Xiao, Yushan Yan

Water electrolysis is critical for the production of green hydrogen for use as a chemical feedstock, transportation fuel, and long-term energy storage solution. However, current electrolyzer technologies, such as alkaline water electrolyzers (AELs), are relatively inefficient due to high internal resistances caused by liquid electrolytes. Proton-exchange membrane electrolyzers (PEMELs), on the other hand, use very efficient solid polymer electrolyte membranes to achieve high hydrogen production rates, but they generate a harsh acidic environment that requires the use of expensive precious metal materials. In this work, we study hydroxide exchange membrane electrolyzers (HEMELs), which use a hydroxide-conducting polymer membrane in an alkaline environment and have the potential to produce hydrogen at high efficiencies and low cost.

In a typical HEMEL, water is fed to the anode side of the cell, while the cathode inlet is kept dry to allow for the easy separation and pressurization of the desired hydrogen product formed by the hydrogen evolution reaction (HER). For the HER to proceed, water must be consumed, and the rate of consumption increases at high operating current densities. As a result, some researchers have reported detrimental "cathode dryout" conditions at high current densities,[1] where transport limitations dominate and prevent any further hydrogen production. In this work, we study the water balance of a HEMEL cathode to improve water transport and delay cathode dryout conditions to higher current densities.

First, we attempt to increase the diffusion of water into the cathode through the membrane by using thinner membranes to decrease the diffusion length scale. We also use different chemistries of membranes with different water diffusivities to make water accessible to the cathode. We then design catalyst layers that can effectively utilize and retain water at the cathode through the use of hydrophilic and hydrophobic additives that alter the water transport pathways through the catalyst-ionomer network. We find that careful tuning of the catalyst layer and porous transport layer wettability can allow for improved utilization of water and easy removal of gaseous hydrogen products to improve HEMEL performance at high current densities. This work could allow for green hydrogen production at improved reaction rates with reduced transport limitations.

(1) Koch, S.; Disch, J.; Kilian, S. K.; Han, Y.; Metzler, L.; Tengattini, A.; Helfen, L.; Schulz, M.; Breitwieser, M.; Vierrath, S. Water Management in Anion-Exchange Membrane Water Electrolyzers under Dry Cathode Operation. *RSC Adv* **2022**, *12* (32), 20778–20784. https://doi.org/10.1039/D2RA03846C.

115f Market Optimization and Technoeconomic Analysis of Hydrogen-Electricity Coproduction Systems

Alexander Dowling

Decarbonization efforts have prompted a rise in variable renewable energy sources, such as wind and solar, and investigations into alternative fuels, such as hydrogen. Integrated energy systems (IESs) are well suited to provide the needed flexibility to support this paradigm shift in the electric grid. Still, conventional techno-economic modeling paradigms neglect the time-varying dynamic nature of the grid and thus undervalue resource flexibility. This talk presents new optimization-based techno-economic analysis methods developed as part of the DOE-sponsored Institute of Advance Energy Systems (IDAES) and Design Integration and Synthesis Platform to Advance Tightly Coupled Hybrid Energy Systems (DISPATCHES) projects. We start by using surrogate models to rigorously optimize the annual operation of six energy system concepts, including solid oxide fuel cell (SOFC) and solid oxide electrolyzer cells, using 61 annual electricity market scenarios and five hydrogen price scenarios. We find that integrated systems that co-produce electricity and hydrogen --- SOFC + SOEC and reversible solid oxide cell (rSOC) --- are profitable in 85% and the top performer in 74% of considered market scenarios. These findings strongly encourage research investments in SOFC and SOEC technologies, especially flexible co-production systems. This analysis, however, relies on the price-taker assumption. In the second half of the talk, we present new multiscale optimization methods that leverage surrogate models to directly embed the impact of IES operation on wholesale electricity market outcomes (e.g., unit commitment and prices). We demonstrate the new approach using two case studies that integrate renewable and nuclear energy sources with electrolyzers to produce hydrogen.

Part 1: This work was conducted as part of the Institute for the Design of Advanced Energy Systems (IDAES) with support through the Simulation-Based Engineering, Crosscutting Research Program, the Solid Oxide Fuel Cell Program, and the Transformative Power Generation Program within the U.S. Department of Energy's Office of Fossil Energy and Carbon Management. Part 2: This work was conducted as part of the Design Integration and Synthesis Platform to Advance Tightly Coupled Hybrid Energy Systems (DISPATCHES) project through the Grid Modernization Lab Consortium with funding from the U.S. Department of Energy's Office of Fossil Energy and Carbon Management, Office of Nuclear Energy, and Hydrogen and Fuel Cell Technologies Office.

205a Thermally and Chemically Stable Metallic Membrane Reactors for H_2 generation with CO_2 Capture

Hani Hawa, Kyle Hawley, Damian Agi, Alexander Dowling

Precision Combustion, Inc. (PCI) is developing a novel membrane reactor (MR) for high efficiency and cost-effective H2 generation and separation with simultaneous CO2 capture, suitable for integration with modular, fuel-flexible gasifiers and reformers.

A proprietary Pd-alloy with tailored composition was identified and demonstrated to enhance the thermal and chemical stabilities of the membrane when operated with simulated water-gas-shift (WGS) conditions. An ultra-thin (<5 μm), defect-free film was deposited onto porous ceramic supports for maximizing flux. Unlike performance reported for prior field demonstrations, the membrane demonstrated high H2 permeance and recovery of up to 90% over 1,200 hours of testing with simulated WGS conditions. The membrane was subjected to multi thermal and pressure cycles and maintained a H2 purity of >99.9%, suggesting no leak growth or H2 embrittlement after cycling. The observed stability of the membrane was correlated to the membrane composition and improved microstructure/synthesis.

Using the Department of Energy's (DOE's) Institute of Design and Advanced Energy Systems (IDAES) resources, we validated a WGS-MR model in Pyomo. By combining the Pyomo model with the H2A model developed by the National Renewable Energy Laboratory (NREL), we performed a preliminary design optimization and technoeconomic analysis (TEA), finding our WGS-MR improved the conversion of feed to H2 by ~22%, resulting in ~14% reduction in levelized cost of H2 (LCOH). We then quantified the impact of the enhanced membrane material properties and identified process integration opportunities to further reduce the LCOH to ~$2/kg H2 (with CO2 capture), ~24% reduction in LCOH vs. the baseline cost.

In future work, PCI will mature the WGS-MR to confirm technical and economic viability of producing high purity H2 from gasified biomass/coal blends. We will also address technology commercialization barriers to transition the technology to fielded applications.

205b Tandem Catalytic Reactor for Single-Step Production of Value-Added Hydrocarbons from Biogas

Olusola Johnson, Babu Joseph, John Kuhn

Biogas, generated through anaerobic digestion of organic matter, is the third largest source of anthropogenic methane emissions in the United States. Typically, biogas energy is either wasted or converted into low-value products, such as heat and electricity. Utilizing biogas-to-liquid technology (BGTL) has the potential to supply 10% of the US diesel market. However, conventional gas-to-liquid (GTL) processes demand significant capital investments and are challenging to scale down. With current crude oil prices, GTL processes struggle to achieve profitability, necessitating simpler and more cost-effective alternatives. Developing affordable small-scale BGTL processes can create new opportunities, such as directly producing liquid fuels from landfill gases.

A key challenge in GTL processes is the need for separate high-temperature, low-pressure reforming and low-temperature, high-pressure liquefaction sections. By developing catalytic systems that merge these two steps into a single reactor, significant cost savings can be achieved. This study presents a novel biofuel production pathway—an intensified process for converting biogas (mainly methane and carbon dioxide) into liquid hydrocarbon fuels in a single reactor. A low-temperature reforming catalyst is combined with a high-temperature Fischer-Tropsch synthesis (FTS) catalyst to consolidate the two reaction steps. The tandem catalytic reactor system employs methane bi-reforming (steam and dry) to generate syngas, which then undergoes C-C coupling through FT catalysis to produce distillate-range hydrocarbon fuels. By combining Ni_3Zn/SiO_2 and Fe_5k_2/SiO_2 catalysts for methane reforming and FTS, respectively, a CH_4 and CO_2 conversion of 79% and 28% was achieved, with stability over 10 hours on stream. The gas phase C_{2+} mass yield was approximately 10%, with a 4.8% C_2-C_4 olefin yield and a 2% C_{5+} hydrocarbon yield based on the feed biogas. This integrated approach demonstrates the potential for an economically viable, single-reactor biogas-to-liquid conversion process.

205c Hydrogen Production from Municipal Solid Waste

Srinivas Palanki

Currently, disposal of municipal solid waste represents a significant societal problem. The increasing piles of waste in landfills pose significant hazards in terms of toxins, leachates, and greenhouse gases. One solution is to utilize the solid wastes as fuel to produce higher-value products. Solid fuels can be used to produce hydrogen via gasification to produce syngas using air or steam. Syngas produced by gasification is converted to CO_2 and H_2 via water gas shift reaction followed by CO_2 and H_2 separation for production of pure H_2. When air is used, the syngas produced is diluted with N_2 lowering the quality of the syngas while contributing to CO_2 emissions. When steam is used as the gasification medium in biomass gasification heat integration schemes such as use of inert heat carriers must be employed to provide heat for the endothermic gasification reaction. Oxygen produced by air separation is traditionally used for oxy-fueled coal gasification. On industrial scale, large amounts of oxygen are required and are typically produced by cryogenic separation from air, a process which requires a large CO_2 footprint, as well as significant capital and operating costs. Thus, the conventional process to utilize solid waste as raw material to produce high-value chemicals poses significant energy and environmental challenges.

Recently, researchers at the National Energy Technologies Laboratory (NETL)[1],[2] have developed novel oxygen carriers for a three-reactor system with promising results using various solid fuels such as biomass, flax straw and MSW. They have also developed a new process to convert CO_2 to CO, allowing for the production of secondary products such as methanol. The main innovation is the development of a novel oxygen carrier process, which eliminates expensive air separation steps, water gas shift reactor and CO_2/H_2separation steps. The process does not require an external heat source eliminating heat integration issues and associated CO_2 emissions. An additional innovation is the fact that the oxygen carrier process operates below 900°C, thereby eliminating the ash melting issues encountered in traditional gasification systems that operate above 900 °C. In this research, NETL's bench-scale experimental results will be utilized to develop a simulation of a large-scale plant in the ASPENPlus environment that utilizes MSW as feed and produces H_2 as the primary product and methanol as a secondary product. This simulation will be utilized to conduct a techno-economic analysis of the entire process. The anticipated advantages of this novel process include lower H_2production costs and the utilization of waste materials as feed such as MSW instead of fossil fuels, resulting in a significant reduction in MSW quantities for disposal and increasing the capacity of the existing waste management infrastructure.

[1] R.V.Siriwardane and J. Riley, "Process for production of hydrogen with transport from chemical looping using mixture of partial oxidation oxygen carriers and combustion oxygen carriers", S 166,257 filed Jan 12, 2022

[2] R. V. Siriwardane and Y. Fan, "Metal Ferrite Oxygen Carriers For Gasification Of Solid Carbonaceous Fuel," U.S. Patent 10,030,204, July 24, 2018

205d Hydrogen Liquefaction: An Overview of the Fundamental Physics, Engineering Practice, Economic Viability, and Future Opportunities

Saif Al Ghafri, Michael L. Johns, Eric F. May

Hydrogen is emerging as one of the most promising energy carriers for a decarbonized global energy system. Transportation and storage of hydrogen are critical to its large-scale adoption and to these ends liquid hydrogen is being widely considered. The liquefaction and storage processes must, however, be both safe and efficient for liquid hydrogen to be viable as such an energy carrier. Identifying the most promising liquefaction processes and associated transport and storage technologies is therefore crucial; these need to be considered in terms of a range of interconnected parameters ranging from energy consumption and appropriate materials usage to considerations of unique liquid hydrogen physics (in the form of ortho-para hydrogen conversion) and boil-off gas handling. This study presents the current state of liquid hydrogen technology across the entire value chain whilst detailing both the relevant underpinning science (e.g. the quantum behaviour of hydrogen at cryogenic temperatures) and current liquefaction process routes including relevant unit operation design and efficiency. Cognisant of the challenges associated with a projected hydrogen liquefaction plant capacity scale-up from the current 32 tonnes per day to greater than 100 tonnes per day to meet projected hydrogen demand, this study also reflects on the next-generation liquid-hydrogen technologies and the scientific research and development priorities needed to enable them. Specific examples of the research on liquid hydrogen currently conducted in Australia will be presented. This study also provides a comprehensive techno-economic assessment demonstrating the viability of a conceptual Australia-Japan liquid hydrogen supply chain. Viability for this specific scenario can be used to indicate broader viability for both Asia-Pacific hydrogen supply chains and potentially hydrogen supply chains globally.

Furthermore, this work will present the design, construction and comissioning of a unique laboratory capability at the University of Western Australia (UWA) that has been recently developed (named LH2Facility) and consists of two major sections: Ortho-para (O-P) hydrogen conversion and hydrogen liquefier and LH2 boil-off monitoring. A single-stage cryocooler is deployed to enable ortho-para measurements at various amounts of catalyst, temperatures ranging between (350 and 30) K, and pressure up to 5 MPa, where ortho-para ratio is monitored by in-situ Raman spectroscopy using in-line fibre optic probe. A dual-stage cryocooler is used to liquefy hydrogen (at around 20 K and 1 bar), which is then transferred to a BOG test cell, where temperature stratification and boil-off rates is studied at different amounts of heat ingress, liquid level, and system pressure. The BOG testing rig has been fabricated, which includes vacuum shields (aluminum and copper cans) and LH2 stainless steel cell. Detailed thermal analyses were performed to minimize heat ingress into the hydrogen liquefier and BOG testing cell. The results were directly applied to the material selection, vacuum considerations, cooling capacity, instrumentation, and the dimension of every component. Overall, the LH2Facility design temperature is 20-350 K with the maximum pressure and volume of 5 MPa and 7 ml for (i) o-p conversion cell; and (ii) 1 MPa and 7 L for the boil-off testing cell. The design, engineering drawings, HAZOP, safety review, SOP, and construction of the LH2Facility have been finalized, and the apparatus is currently being commissioned and tested with liquid nitrogen and liquid neon.

205f Clean Production of Hydrogen from Steam Electrolysis in Molten Salts – Experimental Investigation and Process Modeling

Edward Graham, Bryan Tang, Rui Zheng, Dharik Mallapragada, Yogesh Surendranath

The transition towards a renewable energy-dominated grid has increased the demand for clean and sustainable methods of hydrogen production. Electrolysis is one of the most promising methods for hydrogen production, and this talk focuses on molten salt electrolysis, a relatively new and unconventional approach that allows for steam electrolysis at higher temperatures (>100 C) compared to traditional low-temperature electrolysis systems. This approach is potentially appealing because of many reasons including: a) thermodynamic advantages of reducing electricity input over heat input with increasing reaction temperature, b) likelihood of lower overpotential due to faster reaction kinetics at higher temperatures and c) avoiding reliance on precious materials for catalyzing the reaction. Here, we discuss results pertaining to a) experimental investigation of the proposed approach, b) process model development, and c) subsequent techno-economic analysis and optimization to benchmark the process concept against other low-carbon hydrogen production routes.

The experiments provide insights into the feasibility and practical limitations of the molten salt electrolyser. Cyclic voltammetry experiments have been successfully conducted to determine the polarization curve (voltage-current density dependence) at different experimental conditions and point to the possibility of lower overpotentials compared to state-of-art alkaline electrolysis systems.

Based on these experiments we developed a process simulation model for a production-scale facility that incorporates a custom model for the electrolyser. The model considers various parameters, including the operating temperature and current density, and enables the investigation of the impact of different operating conditions on the efficiency of the process. Additionally, the process model provides a useful tool for identifying how various process design choices, like reaction temperature, heat recuperation mechanisms and cell ohmic potential impact the overall energy requirements of the system and subsequently levelized cost of hydrogen produced.

Techno-economic analyses of the proposed process are performed and results are benchmarked against other approaches to hydrogen production: alkaline, proton-exchange membrane, and natural gas reforming. Due to the highly volatile nature of electricity prices in future renewable energy-dominated grids, both static and dynamic operation of the plant is considered in the techno-economic analyses. The techno-economic modeling results reveal key performance targets for molten electrolysis to be cost-competitive with other approaches for low-carbon hydrogen production.

279a A Decision-Making Framework for Transoceanic Hydrogen Vector Supply Chain Design

Kaveh Khalilpour, Muhammad Aadil Rasool, Ahmad Rafiee, Iftekar Karimi, Reinhard Madlener

Over the last five decades, there have been several phases of interest in the so-called hydrogen economy, stemming from the need for either energy security enhancement or climate change mitigation. None of these phases has been successful in terms of a major market development, mainly due to the lack of cost competitiveness and partially due to technology readiness challenges. Nevertheless, a new phase has begun very recently, which despite holding original objectives has the new motivation to be fully green, based on renewable energy. This new movement has already initiated bipartisan cooperation of some energy importing countries and those with abundant renewable energy resources and supporting infrastructure. One key challenge in this context is the diversity of pathways for the (national and international) export of non-electricity renewable energy. This poses another challenge, i.e., the need for an agnostic tool for comparing various supply chain pathways fairly while considering various techno-economic factors such as renewable energy sources, hydrogen production and conversion technologies, transport, and destination markets, along with all associated uncertainties.

This presentation addresses the above challenge by introducing a probabilistic decision analysis cycle methodology for evaluating various renewable energy supply chain pathways based on the hydrogen vector. The decision support tool is generic and can accommodate any kind of renewable chemical and fuel supply chain option. As a case study, we have investigated eight supply chain options composed of two electrolysers (alkaline and membrane) and four carrier options (compressed hydrogen, liquefied hydrogen, methanol, and ammonia) for export from Australian ports to three destinations in Singapore, Japan, and Germany. The results clearly show the complexity of decision making induced by multiple factors, and that the preferred supply chain combination (electrolyser technology, green energy carrier) in terms of least cost strongly depends on whether the expected levelized cost of hydrogen (ELCOH) or the expected levelized cost of energy (ELCOE) is used as a decision criterion.

279b Feasibility of Importing Solar-Based Green Hydrogen from Australia in the Republic of Korea in Aspects of Both Economic and Environmental Impact

Ayeon Kim, Changgwon Choe, Hankwon Lim

As one of the feasible alternative energy sources, hydrogen(H_2) has been receiving constant attention. However, since H_2 is a secondary energy source produced using primary resources such as natural gas and renewable power, the production capacity of H_2 varies considerably depending on the capable resource amount. Thus, several countries that have insufficient resources may not be able to fulfill their energy demands with their own production and even cause high expenditures for production, requiring importation from overseas. The Republic of Korea is one of the representative countries and has been emphasizing the need to introduce overseas H_2 and striving to promote related projects. In line with the trend, this study aims to verify the competitiveness and feasibility of importing clean H_2 from Australia as a representative case. For the clean H_2 production way, solar energy and water electrolysis basis green H_2 production are considered. Several realistic projects underway between the Republic of Korea and Australia are considered to investigate the possible scope of the case study. By comparing the overseas green H_2 production pathway with the domestic one, the feasibility of importing H_2 in aspects of both economic and environmental impact can be indicated suggesting appropriate guidelines for government and industry to prepare H_2 society by establishing concrete plans to trade H_2 internationally.

279c Strategic Planning of Hydrogen Supply Chains for Meeting Local, Regional, and Global Demands

Kazi Khoda, Akhilesh Gandhi, Faruque Hasan

Decarbonization and reduction of greenhouse gas (GHG) emissions will most likely require significant changes in the energy infrastructure [1]. Hydrogen (H_2) is a versatile fuel, feedstock, and energy carrier that shows a great potential to decentralize our decarbonization efforts towards achieving an equitable low-carbon development [2]. H_2 as a fuel is environmentally benign and being a high-energy dense carrier, it makes energy transportation more economical. Different production routes have been developed for H_2 manufacturing and these have been named with different color names (e.g., blue, green) based on their emissions. Of all the different production routes, green hydrogen is one of the most promising due to its complete dependence on solar and wind power thus having net zero emissions. Availability of renewable energy in some regions makes locally produced *green hydrogen* very attractive. The 2021 Bipartisan Infrastructure Law (BIL) calls for significant investments in American Infrastructure, with over 8 billion dollars in regional clean hydrogen hubs, demonstrating a national commitment and vision towards transitioning to a sustainable hydrogen-based economy (e.g., H2@Scale or Energy Earthshot focusing Hydrogen) [3,4]. The bill also specifically calls for investments in regional hydrogen infrastructure while actively engaging nearby communities. However, it is not clearly understood what fundamental changes in the configurations of future hydrogen infrastructure would lead to a sustainable and equitable carbon-neutral economy. While decentralized manufacturing enables the use of local resources to meet local demands, several barriers need to be overcome before a fully sustainable regional hydrogen economy can be realized. Several barriers exist, including high cost, lack of system-wide integration, safety concerns, regulatory barriers, negative public perception, and gaps in skilled workforce training to participate in a distributed hydrogen economy.

In this study, we present a strategic hierarchical clustering approach for the hydrogen supply chain to satisfy local, regional, and global hydrogen demand. For each location on the geo-spatial grid, we develop a mixed-integer non-linear programming (MINLP) model for the simultaneous design and scheduling of H_2. Essentially, the model optimizes the cost and reports the optimal size of the solar farm, wind farm, electrolyzer, H_2 storage capacity and the cost of H_2. High-fidelity models are developed for the electrolyzer and then reduced order models are implemented in the optimization model. With such a high spatial resolution, data procurement, curation, and analysis is a major challenge for such a large-scale problem [5]. We focus on Texas as a case study and consider the hydrogen production potential from wind and solar resources at the county level, the serviceable consumption potential for hydrogen in the industrial and transportation sectors, and storage, which leads to county-wise local demand. To determine the ideal location for the hydrogen hub/hubs in Texas, we consider the hydrogen production, transportation, and storage cost. We formulate the supply chain element cost function for transportation and storage, including the transformation cost, and analyze centralized vs. distributed hydrogen supply chain. We then conduct clustering to determine the optimal hydrogen supply chain to satisfy the state's country-wise local hydrogen demand economically. Finally, we extend the same hierarchical clustering approach for national, regional, and global levels. This method can assist in locating the optimal location for a hydrogen production facility considering the variable local energy sources and supply chain costs. It can be extended to all states in the United States and other countries with similar energy resources, providing strategic supply chain development and policy reformation suggestions.

References

1. Hasan MMF, Zantye MS, Kazi M-K: **Challenges and opportunities in carbon capture, utilization and storage: A process systems engineering perspective**. *Computers & Chemical Engineering* 2022, **166**:107925.
2. Kazi M-K, Eljack F, El-Halwagi MM, Haouari M: **Green hydrogen for industrial sector**

decarbonization: Costs and impacts on hydrogen economy in qatar. *Computers & Chemical Engineering* 2020:107144.
3. US DoE: **Hydrogen Strategy: Enabling a Low-Carbon Economy**. *Office of Fossil Energy* 2020, **https://www.energy.gov/sites/prod/files/2020/07/f76/USDOE_FE_Hydrogen_Strategy_July2020.pdf**.
4. The White House: **A Guidebook to the Bipartisan Infrastructure Law for State, Local, Tribal, and Territorial Governments, and other partners**. *The White House* 2022, **https://www.whitehouse.gov/wp-content/uploads/2022/01/BUILDING-A-BETTER-AMERICA_FINAL.pdf**
5. Arora A, Zantye MS, Hasan MMF: **Sustainable hydrogen manufacturing via renewable-integrated intensified process for refueling stations**. *Applied Energy* 2022, **311**:118667.

279d A Dynamic Model for Designing Optimal Condition of the Tank on Hydrogen Carrier

Taejong Yu, Youngsub Lim

This study presents a dynamic model developed to analyze the storage capacity of hydrogen, holding time, and hydrogen loss depending on storage pressure of the tank. The escalating concern regarding climate change and the growing demand for environmentally sustainable energy has led to an increasing interest in hydrogen and hydrogen carrier to transport hydrogen by the ship. We consider the cryogenic pressure tank that is initially filled with 99.81% para hydrogen, whose temperature and pressure increase and which is brought to supercritical condition by external heat ingress. The ortho-para equilibrium of hydrogen is primarily determined by temperature; however, experimental evidence has demonstrated that the conversion rate of para-to-ortho hydrogen is influenced by density and temperature. The conversion rate of endothermic para-to-ortho conversion and absorbed heat are affected by the initial pressure, amount of hydrogen, temperature and external heat ingress. These variables have a significant impact on the duration of time that the tank can hold hydrogen before it reaches its maximum allowable pressure and the amount of boil-off gas loss to maintain the tank pressure. Based on this model, we aim to derive optimal storage condition for 10K hydrogen carrier during its voyage considering the amount of stored hydrogen, external heat, and boil-off gas loss.

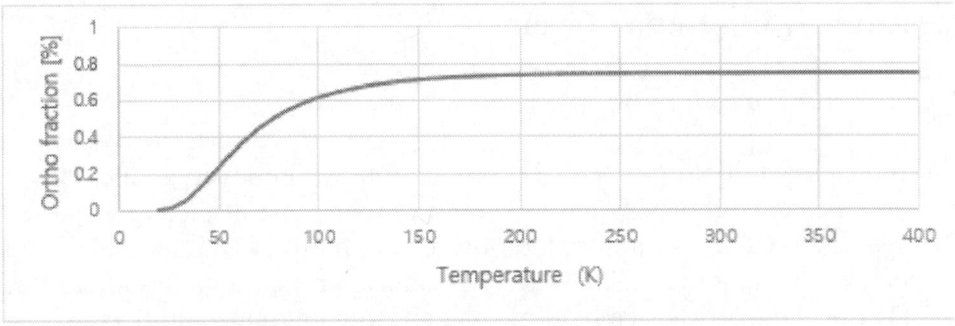

Table 1. Ortho-Para equilibrium depending on temperature

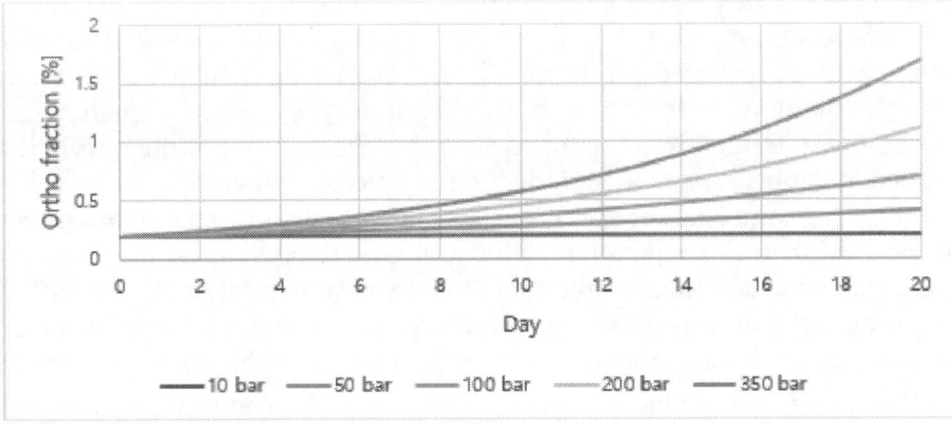

Table 2. Ortho hydrogen fraction during voyage varying with the pressure

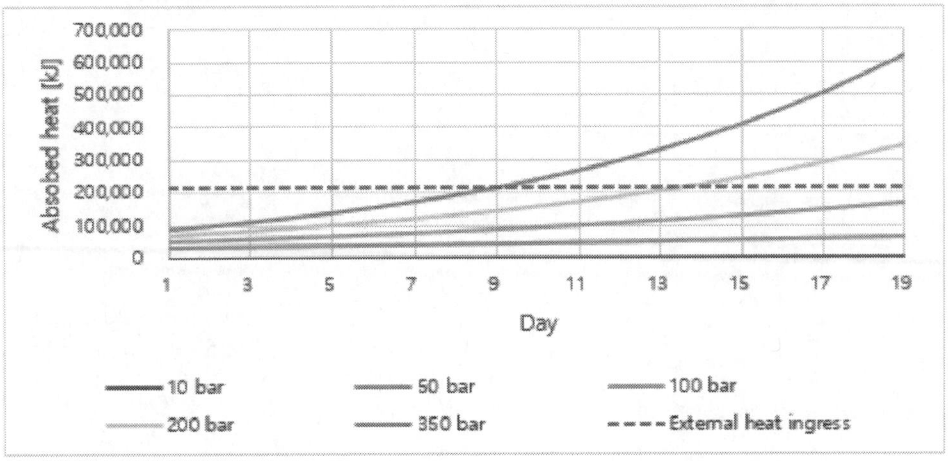

Table 3. Absorbed heat by the para-to-ortho conversion varying with the pressure.

279e Solar-Thermal Processing to Split Water and/or Carbon Dioxide Via a Near-Isothermal, Pressure-Swing Redox Cycle

Kent J. Warren, Liam Taylor, Justin Tran, Alan Weimer

Technologies like solar thermochemical hydrogen (STCH) production, which leverage the oxygen-exchange properties of metal oxides to split water (and/or carbon dioxide), are attractive because they can harness the entire solar spectrum, whereas photovoltaic-driven alternatives (e.g., water electrolysis) rely only on specific wavelength ranges. Generally, this approach involves alternating between two steps namely, (1) the endothermic removal of oxygen from an oxide and (2) the exothermic oxidation of the reduced oxide with water to produce hydrogen. The endo/exothermic nature of each reaction implies that opposite changes in temperature are required to drive the hydrogen yield per cycle. However, in practice, implementing a temperature swing between redox regimes imposes significant irreversibilities that arise from sensible heating of the solid, thus resulting in inefficient operation unless heat is recovered via ultra-high temperature thermal energy storage (a technology still in its infancy). Therefore, in this work, a scalable prototype reactor – situated within a well-insulated cavity receiver – was designed and operated such that the water-splitting (oxidation) step was intentionally initiated at temperatures where reduction is favorable, thereby avoiding the complications associated with solid-solid heat recuperation. To mitigate the undesired thermodynamic consequence of performing the exothermic oxidation reaction at higher temperatures, iron aluminates, a low-cost class of materials with an exceptional isothermal (and near-isothermal) capacity for hydrogen production, were considered for this prototype demonstration. The stable experimental results, obtained under concentrated radiation from a 45 kW_e high-flux solar simulator, provide compelling motivation for the commercial viability of solar thermochemical routes for hydrogen production.

279f Operational, Economic, and Environmental Comparisons of Solar Energy Integration Methods for a High Temperature Steam Electrolysis Plant

Jake Immonen, Kody Powell

There has been an increasing push to produce hydrogen in a low carbon output manner for both its current main use as a chemical feedstock and its promising new uses as an energy storage carrier. One method to produce hydrogen is through high temperature steam electrolysis (HTSE). HTSE utilizes steam at temperatures around 800 °C to produce a hydrogen and oxygen product by inducing a current through solid oxide electrolysis cells (SOECs). One of the main benefits of HTSE is that the electrical input to the electrolysis reaction is reduced by thermal heating to produce steam. In addition, this thermal input can be significantly lowered through efficient heat recuperation from the electrolyzer product streams. There have been large efforts to commercialize SOEC technology to build large scale HTSE plants but there are many questions that need to be addressed to find the most efficient designs and best ways to incorporate renewable energy to make clean hydrogen.

This study focuses on the analysis of different feasible solar energy integration methods for a solar-to-hydrogen high temperature steam electrolysis (SH-HTSE) plant. Three solar energy integration scenarios and one baseline, Grid Only scenario are analyzed as shown in the Abstract Figure. The scenarios are compared using operational, economic, and environmental metrics. Two scenarios, the PV Design and Mixed PV and Solar Thermal Design (Mixed Design), only use renewable electricity from solar PV while producing hydrogen, limiting their production to daytime. The Mixed Design supplements the heat requirement for steam generation with solar thermal energy while the PV Design uses electric resistive heating for steam generation. Despite using resistive heating for steam generation, the PV Design is still more electrically efficient than proton exchange membrane electrolysis due to efficient heat recuperation from the product streams. The Solar Thermal Design and Grid Only Design produce hydrogen continuously using electricity provided from the grid. When possible, the Solar Thermal Design uses solar thermal heat for stream generation, and otherwise uses electric resistive heating. Results show that if tax credits from the Inflation Reduction Act are included, the PV Design is the most economical choice with a levelized cost of hydrogen value close to \$3/kg H_2 produced. If tax credits aren't included the Solar Thermal Design is the most economical choice but this is at the expense of carbon emissions with both the Solar Thermal Design and Grid Only Design outputting more carbon emissions than steam methane reforming due to the current carbon heavy nature of the electrical grid. The main downside of the PV and Mixed Design is that they will have to ramp production to meet the electric generation from PV. This operation is investigated through dynamic modeling of the SOECs from hot standby mode to full production to show how a multi-megawatt electrolysis system could follow generation without compromising the SOECs' health. Overall conclusions from this work highlight that using PV for the electric input to the SH-HTSE plant can demonstrate promising operational, economic, and environmental results and should be investigated further for sustainable, clean hydrogen production.

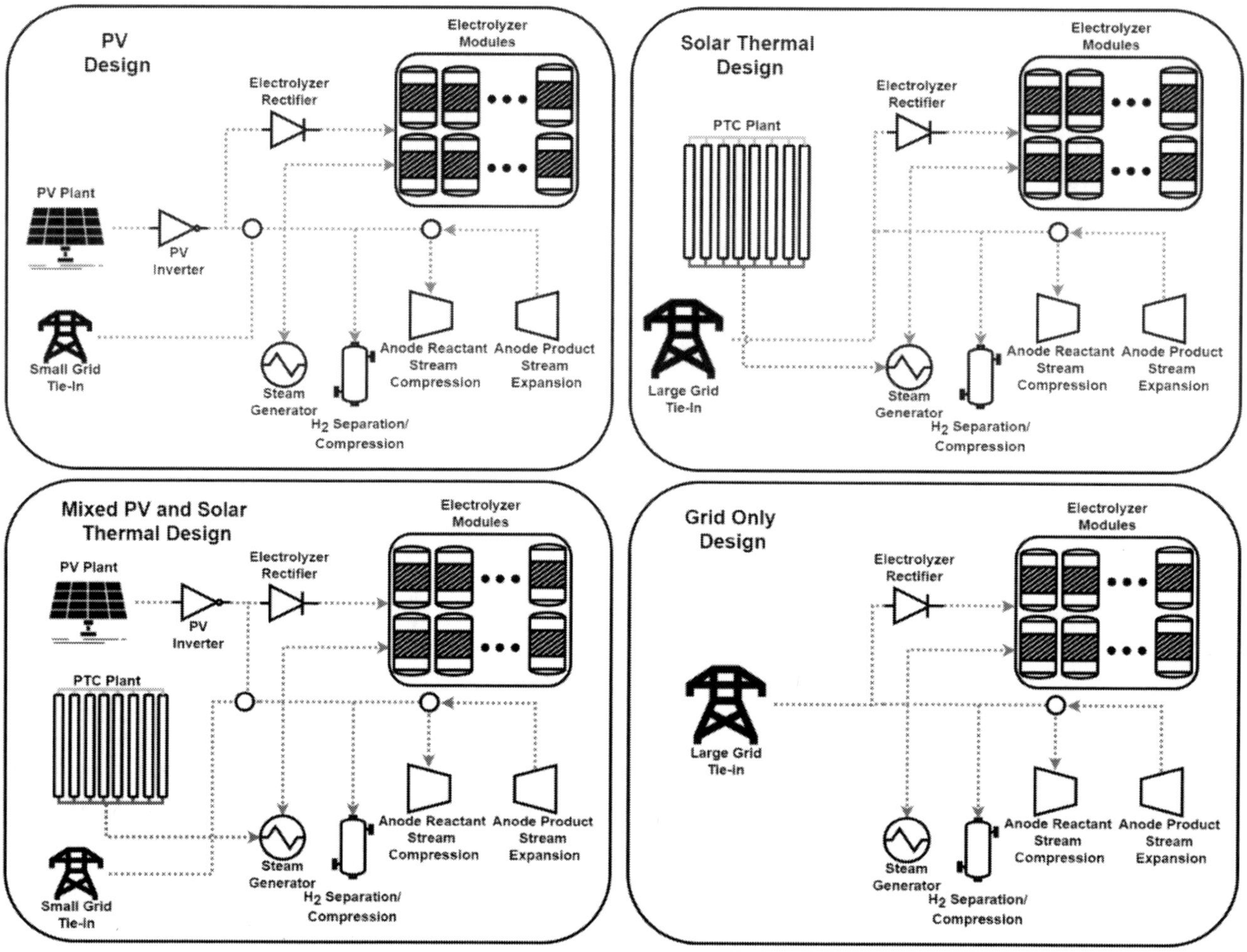

Abstract Figure: Three solar energy integration designs and baseline (Grid Only) design detailed in this study. Purple dashed lines show electricity use to major electricity consuming equipment while red dashed lines show solar thermal heat.

336a UK-Hyres - Identifying, Prioritising and Solving Research Challenges in Hydrogen and Alternative Liquid Fuels

Rachael Rothman

UK-HyRES is the flagship UK research hub for hydrogen and alternative liquid fuels. It aims to bring together the hydrogen community to identify, prioritise and solve challenges that limit deployment of hydrogen and alternative liquid fuels. The hub is focussed around four technical themes: production, storage and distribution, end use, and alternative fuels (predominantly ammonia). All technical work and projects are looked at through the cross cutting lenses of economic, environment, social and safety to ensure that technical solutions are sustainable, safe and can deliver impact. This presentation will cover initial projects within the hub, including hydrogen for aviation, the energy networks needed for industrial hydrogen deployment across foundation industries, and green ammonia production.

336b Chemical Looping Single Reactor System for Biogas Conversion to Liquid Fuels Using Modified Calcium-Iron Based Oxygen Carrier Particle

Rushikesh Joshi, Ishani Karki Kudva, Pinak Mohapatra, Shekhar Shinde, Liang-Shih Fan

One of the biggest challenges that the world faces is climate change. According to the IPCC report 2022, the average temperature is set to reach 1.5 degrees Celsius above pre-defined levels, and this demands immediate attention to cut down on fossil fuels and shift towards greener solutions [1]. There are several candidates for alternative energy sources. Biogas, derived from organic matter such as animal or food waste, is one of the promising routes toward a greener future [2].

Studies have shown that the major setback in biogas utilization is its high carbon dioxide content. This drawback, consequently, results in lower energy efficiency and reduced conversion rates. Conventional pathways to upgrade biogas include carbon dioxide removal to produce methane or reforming the biogas to produce syngas via dry reforming or tri reforming process [3]. However, both these processes have high energy requirements, rendering the use of biogas ineffective. Hence, the whole process becomes highly energy intensive. The need of the hour is to design a carbon-neutral and highly efficient process.

This study proposes a chemical looping single reactor system for the conversion of biogas to liquid fuels using a modified Calcium-Iron based oxygen carrier. Biogas contains methane, carbon dioxide, nitrogen, and trace amounts of volatile organic compounds. A single reactor system can be divided into two zones: a) Particle reduction zone and b) Particle regeneration zone. Biogas along with the proposed oxygen carrier, i.e., $Ca_2Fe_2O_5$, is fed from the top of the moving bed reactor and reacts in a co-current fashion. Simultaneously, a stream of steam and carbon dioxide from the bottom reacts with reduced oxygen carrier particles. The product, which is a high-purity syngas, is removed from the middle of the reactor system. The regenerated material is fed back to the reactor through a riser. The selection of oxygen carriers accounts for the higher yields of syngas.

This process is further simulated in Aspen Plus to study the effect of temperature, pressure, gas composition, and oxygen carrier flow rate. Simulations have shown that the oxygen carrier flowrate influences the syngas purity, solids conversion as well as H_2/CO ratio. The highest syngas purity was observed in the feed ratio of 80:20 CH_4 and CO_2 mixture, whereas the solids conversion rate decreases with increasing flowrate. Moreover, the single reactor system can be operated under autothermal conditions with an oxy-combustion burner compensating for the endothermic heat. This has been confirmed through Aspen simulations.

This newly proposed system utilizing a single reactor has demonstrated impressive capability in producing high purity syngas, despite significant variations in the methane to carbon dioxide ratio. This system was compared with the traditional tri-reforming process for carbon dioxide utilization and syngas production for liquid fuels. The results revealed a remarkable improvement of 15-20% in carbon dioxide utilization using the single reactor system as compared to the tri-reforming process.

To further improve the oxygen carrier, Studies were conducted using Density Functional Theory (DFT) to find a suitable dopant with a strong affinity toward regeneration kinetics. Cobalt was found to increase the reactivity of carbon dioxide in oxidation, allowing the oxygen carrier to be regenerated entirely in CO_2/H_2O in the single reactor system [4]. Samples with different Cobalt compositions were prepared and tested in a Thermogravimetric analyzer for their reactivity and recyclability, aligning well with XRD and SEM analysis on the doped oxygen carrier.

References:

[1]D. R. M. T. E. P. K. M. A. A. M. C. S. L. H.-O. Pörtner, 'IPCC, 2022: Climate Change 2022: Impacts, Adaptation and Vulnerability,' Cambridge University Press. Cambridge University Press, Cambridge, UK and New York, NY, USA,, New York, NY, USA,, 2022.

[2]A. N. L. S. V. S. I. S. Vladimir Arutyunov, 'Utilization of renewable sources of biogas for small-scale production of liquid fuels,' *Catalysis Today,* 2020.

[3]B. J. S. XianhuiZhao, 'Biogas Reforming to Syngas: A Review,' *iScience,* 2020.

[4]Z. C. D. S. B. J. A. F. L.-S. F. Vedant Shah, 'Highly Selective Production of Syngas from Chemical Looping Reforming of Methane with CO2 utilization on MgO- Supported Calcium Ferrite Redox material,' *Applied Energy,* 2020.

336c Sustainable Aviation Fuel Production: An Enviro-Economic Assessment of Direct CO_2 Hydrogenation

Andrea Bernardi, Daniel Bagan Casan, Andrew Symes, Benoit Chachuat

Reducing the greenhouse gases emissions from human activities is a key challenge towards net-zero and a sustainable future (IEA 2020). Transportation plays an important role in everyday life and the aviation industry emits 2% of the total GHG emissions, with a consumption of 10% of the fuel worldwide, and is recognised as a particularly hard to decarbonise sector. Several challenges reduce the number of alternatives to fossil fuels for the aviation sector, including the high energy content per unit mass and volume required and the long service life of the airplanes. It is expected that a large portion of the current global fleet will continue to be operational until 2040-2050 calling for the development of drop-in alternatives based on sustainable resources (Royal Society 2023). According to IEA (2022) low-carbon hydrogen-derived liquid fuels will gain significant market share in the sector, as batteries and hydrogen are not feasible alternatives, especially for long-rage flights, due to the low energy content per unit of mass and volume respectively (Karadotcheva *et al.* 2021). Moreover, hydrogen-fuelled aircrafts would require a complete redesign and advanced materials to avoid hydrogen embrittlement and leakage (Airbus 2020). Lastly, drop-in fuels can leverage existing infrastructure for fuel distribution and storage, which is an important enabler for a fast transition.

Sustainable aviation fuel (SAF) is a broad term that includes biofuels and synthetically derived hydrocarbons, or e-fuels. Significant research efforts have been devoted to identifying routes to produce liquid fuels from CO_2 and H_2 leveraging the Fisher-Tropsch (FT) synthesis, a well-established thermochemical process to produce hydrocarbons of different chain length distribution from syngas, a mixture of CO and H_2. However, syngas is nowadays produced from steam methane reforming or coal gasification and carries a high environmental burden. To reach a significant reduction in the greenhouse gas emissions, CO_2 (biogenic of from direct air capture, DAC) must be used as the carbon source in the FT process along with sustainable H_2. One option is to convert CO_2 and H_2 in a reverse water gas shift (RWGS) reactor and use the resulting syngas in a traditional FT process (Zang *et al.* 2021). The main limitations of this two-step approach are: (i) RWGS reaction requires high temperatures to achieve a good conversion (600-1000 °C); and (ii) the production of wax, a mixture of heavy hydrocarbons, requires additional upgrading units, such as a hydrocracking (HC) reactor to improve the yield of liquid fuels. The direct hydrogenation of CO_2 to liquid fuels in a single step is more appealing, but the selectivity towards liquid fuels is usually low, and the formation of short chain hydrocarbons is favoured (Gao *et al.* 2017). Recently, Yao *et al.* (2020) synthesized a novel Mn-Fe-K based catalyst capable of converting CO_2 and H_2 with excellent yield and selectivity towards liquid hydrocarbons in the jet fuel range (C8–C16) and minimal wax production.

In this work, we carry out a techno-economic analysis and life-cycle assessment (LCA) of two SAF production processes: a one-step process (1sFT) based on the above-mentioned catalyst; and a two-step process (2sFT) based on the work of Zang et al. (2021). As feedstock, we consider CO_2 from DAC as carbon source, available at 25 °C and 1 bar (Keith *et al.* 2016) and H_2 from water electrolysis using wind energy, available at 20 bar and 25 °C (Al Qahtani *et al.* 2021). Both feed streams are assumed to have no impurities. 1sFT and 2sFT are simulated at scale using Aspen HYSYS to evaluate their economic performances and obtain inventories for the LCA. The environmental assessment is conducted in SIMAPRO 9.3, using Ecoinvent 3.8 Cut-Off database for the background process inventories. The fuel obtained from 1sFT and 2sFT is compared against the fossil-based alternative considering different environmental KPIs, as for SAF to be truly sustainable the reduction is GHG emission should not be associated with high collateral ecological damage, a problem often referred to as burden shifting. ReCiPe2016 is used as life-cycle impact assessment method (Huijbregts *et al.* 2017) and GWP100 is evaluated alongside the monetized endpoint impacts to human health, ecosystem quality and resource scarcity. The monetisation factors to translate the endpoint impacts to a common monetary basis are obtained from Dong *et al.* (2019) converted to $\$_{2021}$.

Our results show that the one-step process is superior both in economic and environmental terms to the two steps process, due to a lower capital cost, higher selectivity towards liquid hydrocarbons and lower energy requirements. On a well-to-tank basis both 1sFT- and 2sFT-derived fuels have a negative GWP potential and outperform the fossil-fuel counterpart, while on a well-to-wake basis the 1sFT process is predicted to reduce GHG emissions by 75% and the 2sFT process by 60% (Figure 1a). The analysis of the endpoint environmental impacts confirms a better environmental performances of the synthetic fuels compared to the fossil counterpart, with the 1sFT outperforming the 2sFT by a larger extent, with 1sFT and 2sFT having a total externalities cost 80% and 30% lower that the fossil fuel, respectively (Figure 1b). The largest share of externalities cost in both the one-step and two-step processes is attributed to the damage to resource scarcity due to the high energy requirements of DAC and therefore alternative DAC processes should be considered to move further away from the fossil-dependency of the aviation sector. At current CO_2 and H_2 prices the productions cost of low-carbon synthetic fuels is predicted to be 4-6 times higher than fossil-based aviation fuels, which is in agreement with the literature. And even if the CO_2 and H_2 prices were to decrease significantly, policy interventions such as carbon taxation would likely remain necessary for synthetic fuels to become cost competitive. Future work will focus on include additional options for H_2 and CO_2 procurement, including blue hydrogen, alternative DAC processes, and biogenic carbon sources.

Acknowledgement: Funding from the Engineering and Physical Sciences Research Council (EPSRC) for the research under the UKRI Interdisciplinary Centre for Circular Chemical Economy programme (EP/V011863/1) is gratefully acknowledged.

References

Airbus, 2020, Airbus reveals new zero-emission concept aircraft. https://tinyurl.com/AirbusNet0 [Accessed 03/04/2023].

Al-Qahtani *et al.* 2021. Uncovering the true cost of hydrogen production routes using life cycle monetisation. Applied Energy, 281, 115958.

Dong *et al.* 2019. Evaluating the monetary values of greenhouse gases emissions in life cycle impact assessment. Journal of Cleaner Production, 209, 538-549.

Gao *et al.* 2017. Direct conversion of CO_2 into liquid fuels with high selectivity over a bifunctional catalyst. Nature Chemistry, 9, 1019.

Huijbregts *et al.* 2017. ReCiPe2016: a harmonised life cycle impact assessment method at midpoint and endpoint level. The International Journal of Life Cycle Assessment, 22(2), 138-147.

IEA 2020, CCUS in Clean Energy Transitions, IEA, Paris https://www.iea.org/reports/ccus-in-clean-energy-transitions, License: CC BY 4.0

IEA 2022, Aviation, IEA, Paris https://www.iea.org/reports/aviation, License: CC BY 4.0

Karadotcheva *et al.* 2021. Structural power performance targets for future electric aircraft. Energies, 14(19), 6006.

Keith *et al.* 2018. A process for capturing CO_2 from the atmosphere. Joule, 2(8), 1573-1594.

The Royal Society, 2023. Net zero aviation fuels – resource requirements and environmental impacts policy briefing.

Yao *et al.* 2020. Transforming carbon dioxide into jet fuel using an organic combustion-synthesized Fe-Mn-K catalyst. Nature Communications, 11(1).

Zang *et al.* 2021. Performance and cost analysis of liquid fuel production from H_2 and CO_2 based on the Fischer-Tropsch process. Journal of CO_2 Utilization, 46.

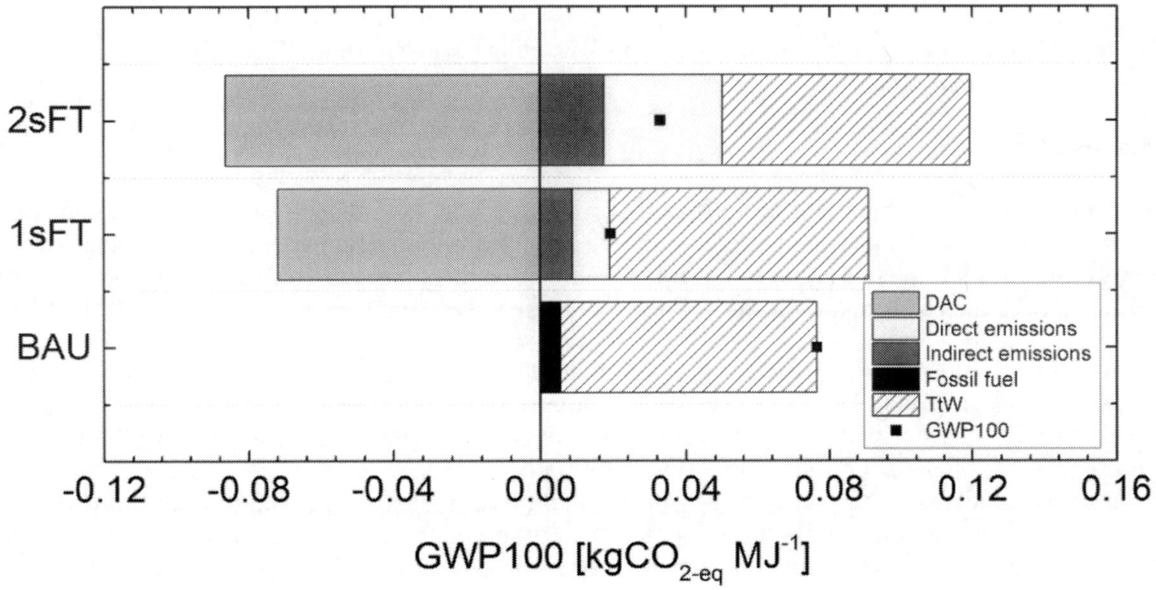

(a)

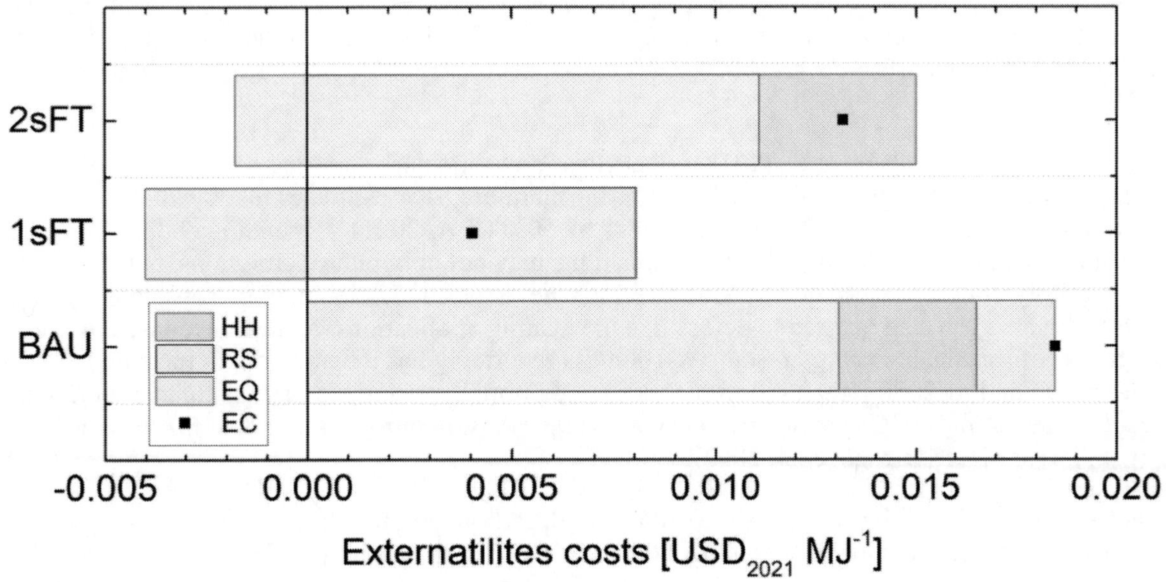

(b)

Figure 1: (a) Well-to-Wake GWP100 for the fossil-derived fuel (business-as-usual, BAU), one-step process (1sFT) and two steps process (2sFT); (b) total externalities cost for BAU, 1sFT and 2sFT processes.

336d Renewable Energy Integration and Waste Heat Recovery for the Production of Sustainable Jet Fuel and Decarbonization of Industrial Heating Applications

Daniel Florez-Orrego, Dareen Dardor, Meire Ellen Gorete Ribeiro Domingos, Reginald Germanier, François Maréchal

Extended abstract

Heat use in industry is responsible for a significant portion of global greenhouse gas emissions, mainly due to the intensive use of natural gas and petroleum derivatives to drive different chemical and industrial processes. However, by transitioning to renewable resources, such as biomass and electricity, the industry can significantly reduce its carbon footprint.

For the steel industry, the use of electric arc furnaces that rely on electricity to melt scrap steel (instead of blast furnaces, which rely on coke) can reduce carbon emissions by up to 80%, (RMI, 2022). In the cement industry, alternative fuels derived from biomass along with carbon capture and utilization (CCU) can be an option to upgrade the biogenic emissions and store them to produce synthetic fuels when cheap electricity from prosumers is available in the summer. As for the aluminium industry, aluminum produced with renewable electricity can reduce emissions by up to 90% compared to conventional aluminum production, according to the report by the International Energy Agency (IEA) and the Aluminium Stewardship Initiative (ASI, 2021). Also, the aviation sector is a significant contributor to greenhouse gas emissions, thus the production of sustainable aviation fuel (SAF) is one approach to reducing emissions. SAF is a drop-in replacement for conventional jet fuel that can be produced from renewable resources such as biomass and electricity. Since the heating requirements of the Fischer Tropsch process can be also supplied by electrification or via implementation of biomass thermal gasification, SAF produced from biomass and electricity could reduce the lifecycle greenhouse gas emissions up to 85%, compared to conventional jet fuel (ICCT, 2022).

According to a study by Jacobson et al. (2018), electrification of industry and transportation sectors together can reduce global carbon emissions by 68%. Furthermore, IEA estimates that by 2050, electrification could reduce industrial sector emissions by 50% (IEA, 2021). Depending on the source of electricity, electrification of industrial heating applications may not only reduce emissions, but also improve the energy conversion efficiency. Moreover, whenever electrification may be considered inapplicable, biomass can also be used to defossilize the heating applications, reduce the emissions and promote the use of sustainable energy resources. Biomass reportedly has the potential to meet up to 40% of global energy demand by 2050 (Hoogwijk et al. 2005). Additionally, using biomass in combination with carbon capture and storage (CCS) may lead to negative emissions, helping to mitigate climate change (Smith et al., 2016; Florez-Orrego, et al. 2019).

The advantages of the use of the biomass gasification for decarbonizing the industrial applications are multiple. First, the waste heat available at about 800 °C-1000°C in the gasifier outlet stream suitably matches the high-grade heat temperature required for preheating or even melting some substances, leading to a reduction of combustion of natural gas. Moreover, the syngas produced is a staple feedstock to produce either synthetic natural gas and hydrogen, which can be stored and consumed to balance the heating requirements of the industrial applications. Methanation reaction for synthetic natural gas production also shows temperatures levels (300°C to 500°C) that fit the range of temperature found in biomass pretreatment processes of even metal preheating, such as sow dryers in aluminium remelting plants. In this regard, both the waste heat and, in less extent, the produced gas, can be a source of heat for decarbonizing the energy systems and also provide anergy heat to the surrounding populations.

On the other hand, if syngas produced is more than the amount required due to the seasonal operation or maintenance stops, syngas can still be used to produce Fischer-Tropsch liquids, which can minimize or even avoid the dependence of fossil fuels of hard-to-defossilize sectors, such as air transport. The temperature levels at the Fisher Tropsch process (including the offgas consumption) may lead to a surplus power generation in organic Rankine cycles and supercritical CO2 cycles, further increasing the overall energy efficiency of the energy conversion systems. This electricity can be in turn consumed in a co-electrolysis system that convert CO2 derived from the gasification process into more CO and H2, which are the feedstock to produce Fischer Tropsch liquids. The higher efficiency of these advanced energy systems can help mitigating the foreseen increment in the biomass- and electricity-derived fuels production, considering that, in 2021, United Airlines announced that it would purchase up to 10 million gallons of SAF from World Energy, making it the largest airline purchaser of SAF in the world. Other airlines, including Delta Air Lines, Lufthansa, and KLM, have also committed to using SAF (Reuters, 2021).

The process synthesis and optimization methods used in this work include simulation of biomass gasification, hydrogen and synthetic natural gas production, and jet fuel production via a Fisher Tropsch unit. Aspen Plus software has been used to calculate the properties of the conventional and non-conventional components, along with the heating, cooling and electricity demands of the energy systems. An energy integration approach, carried out by the OSMOSE platform, allowed minimizing the energy requirement of the integrated systems and prioritizing the use of waste heat at high temperature for industrial applications. In this way, the low grade waste heat can be reutilized to supply a nearby district heating network or a supercritical CO2 cycle to generate surplus power. In addition, the implementation of a seasonal power-to-gas approach, including electrolysis, methanation, and carbon abatement technologies, required the management of the time-varying demand and supply. To this end, a multi-time approach, using seasonal profiles of energy demands and prices, has been used to minimize the capital investment of the seasonal energy storage systems (e.g. liquids tanks). In the attached Figure, an industrial plant is represented as the main energy system to be heated by the use of renewable energy resources, whereas the city operates as the heat sink thereof. Accordingly, the goal was to reduce the valuable waste heat released to environment during the summer period and store it, so that it can be used to supply the energy needs in the winter, along with biomass, to the SAF production plant.

The time-varying demands, along with the number of decision variables required the application of the OSMOSE platform, which solved the MILP problem that worked out the best operating conditions and sizes of the equipment, maximizing the benefits while reducing the wastes. This platform is capable of integrating Aspen Plus and AMPL software in order to automatize the calculations and provide solutions to several scenarios of operation under different market conditions. Thanks to its versatility, the combination of different energy technologies has been applied to defossilize industrial heating applications, whereas producing value-added fuels. Typical energy demands of metal (aluminium remelting process), beverages (brewery process) and food (dairy process) industrial process were used to exemplify the application of this approach. A typical central European zone city is considered for the assumption of the thermal loads of district heating network, including domestic hot water, space heating, air conditioning and refrigeration. The integration with the surrounding population has been achieved implementing a novel CO2 district heating network, which can be used to distribute heat and upgrade the low-grade waste heat released by the industrial production processes making use of heat pumps. As a result, the fossil emissions could be virtually avoided and a production of up to 100000 barrel of synthetic aviation fuel per day has been achieved. Also, the installation of a CO2 management system, composed of liquefaction, storage and recompression units allowed injecting biogenic CO2 (120EUR/t), which translated to higher economic benefits of the integrated energy system.

References

Flórez-Orrego, D. et al (2019). Comparative exergy and economic assessment of fossil and biomass-based routes for ammonia production, Energy Conversion and Management, v.194, pp. 22-36.

Reuters. (2021). United Airlines unveils plan to fund more sustainable jet fuel made from trash. Available at: https://www.reuters.com/business/sustainable-business/united-airlines-unveils-bid-fund-sustainable-aviation-fuel-use-2021-04-13/ .

Rocky Mountain Institute (RMI, 2022). Steel Supply Chain Emissions Transparency: A Product-Level GHG Accounting Approach. Available at: https://rmi.org/wp-content/uploads/2022/07/horizon_zero_steel_two_pager.pdf

Aluminium Stewardship Initiative (ASI). (2021). ASI's position and direction on a 1.5°C pathway aligned aluminium sector. Available at: https://aluminium-stewardship.org/asis-position-and-direction-on-a-1-5c-pathway-aligned-aluminium-sector

International Council on Clean Transportation (ICCT). (2022). Sustainable aviation fuels: what does real leadership look like? Available at : https://theicct.org/publications/renewable-jet-fuel-ghg-lca-2019

Jacobson, M. Z., Delucchi, M. A., Cameron, M. A., & Mathiesen, B. V. (2018). Matching demand with supply at low cost in 139 countries among 20 world regions with 100% intermittent wind, water, and sunlight (WWS) for all purposes. Renewable Energy, 123, 236-248.

International Energy Agency (IEA). (2021). Net Zero by 2050: A Roadmap for the Global Energy Sector. IEA.

Hoogwijk, M., Faaij, A., Eickhout, B., & de Vries, B. (2005). Potential of biomass energy out to 2100, for four IPCC SRES land-use scenarios. Biomass and Bioenergy, 29(4), 225-257.

Smith, P., Davis, S. J., Creutzig, F., Fuss, S., Minx, J., Gabrielle, B., Canadell, J. G et al. (2016). Biophysical and economic limits to negative CO2 emissions. Nature Climate Change, 6, 42-50.

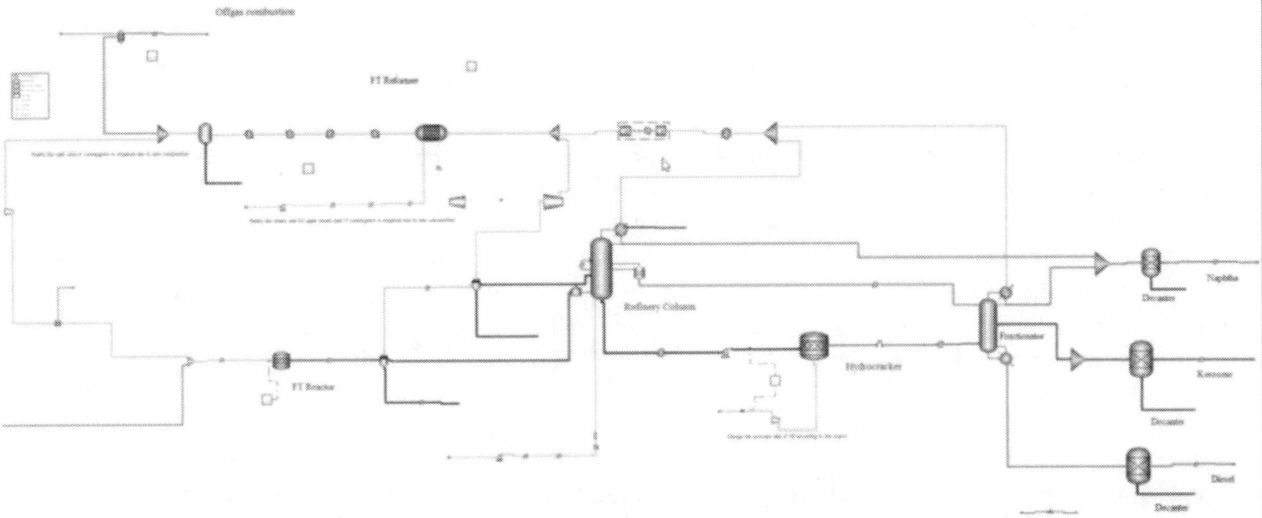
Offgas combustion
FT Refiner
Refinery Column
Hydrocracker
FT Reactor
Fractionator
Naphtha
Kerosene
Diesel
Decanter
Decanter
Decanter

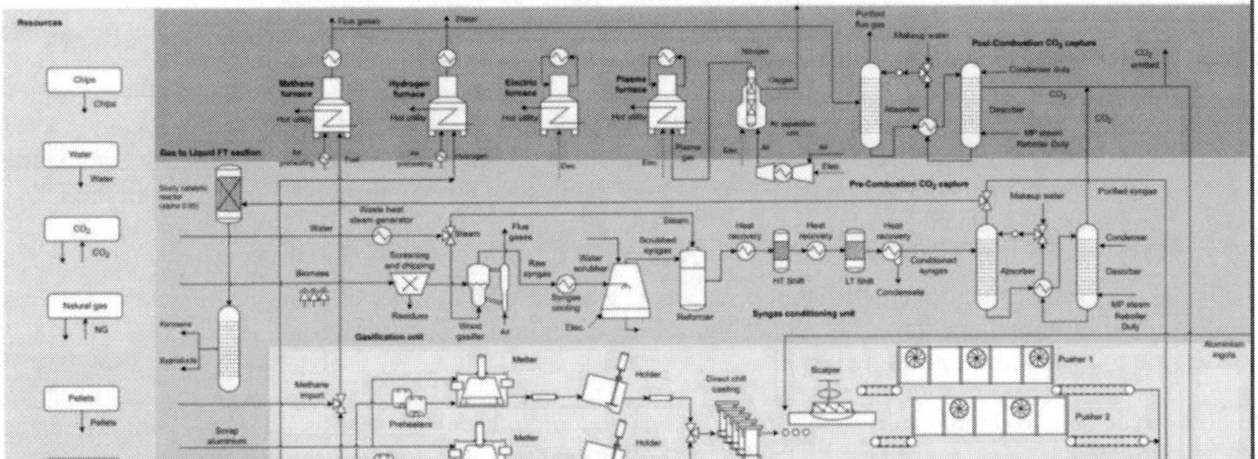

336e Methanol As a Near-Term Solution for Monetizing Renewable Energy and Green Hydrogen.

Maxim Lyubovsky

A sheer scale of human addiction to fossil fuels makes global decarbonization within two or three decades, which is required to prevent extreme consequences of the climate change, an extremely aggressive target. This does not leave much time for development of new power technologies or building new energy infrastructures to replace the ones built over the last two centuries around the use of fossil coal, oil, and natural gas. Production of net-zero carbon hydrocarbon fuels by combining captured CO_2 with green hydrogen produced by water splitting utilizing renewable carbon-free energy sources provides an alternative "shovel ready" de-fossilization solution where renewable fuels compatible with the established energy infrastructures and existing power applications will directly substitute for and gradually replace fuels currently produced from fossil oil, coal, and natural gas. Ultimately this will create the indefinitely sustainable carbon cycle (figure 1) in which CO2 essentially serves as a backbone to which renewable energy is attached in the form of green hydrogen to be used in all existing power applications. No additional extraction of fossil carbon would be required.

Methanol – the simplest alcohol – is a particularly attractive chemical to be used as a large-scale renewable energy carrier. Methanol is stable liquid under wide range of normal ambient conditions. It is already an important commodity chemical with global annual production of about 100 million tonnes and is used in production of 245 different industrial chemicals. Methanol is produced under mild temperature and pressure conditions (about 200-250oC and 30-50 bar), which do not require exotic and expensive alloys for reactor construction, and over inexpensive CuZn based catalysts. Only minor modifications from the existing commercial methanol synthesis process would be required to produce methanol from captured CO_2 and H_2 produced from electrolysis. Well-developed worldwide storage and distribution network for methanol is already established. Methanol is also used in a growing number of energy applications, such as production of biodiesel, methanol blends with gasoline, direct use as automotive and marine fuel, conversion to synthetic aviation fuel, etc.

In order to be able to rapidly substitute for fossil fuels on a global scale cost of production of renewable alternatives should be competitive with the costs of traditional fuels. Cost analysis of a standalone re-methanol plant, which incorporates a wind farm, water electrolysis units, buffering storage for hydrogen and CO2 and a methanol synthesis unit demonstrates that for the near-term case the renewable energy incentives provided by the Inflation Reduction Act reduce the cost of re-methanol production to as low as $232/MT. Under the future case assumptions, the production cost is estimated at $540/MT without incentives. Both costs are lower than the current market prices for grey methanol and are at the lower range of historical prices for other common fuels when normalized to the fuel heating value.

Cost of CO2 remains a significant uncertainty in the cost of re-methanol production. For the near-term production case $50/t_CO2 was used, assuming CO2 capture from industrial point sources. Direct air capture of CO2 (DAC) at $150/t_CO2 was assumed for the future case. Recent advancements in development of DAC technologies indicate that this cost assumption is very reasonable.

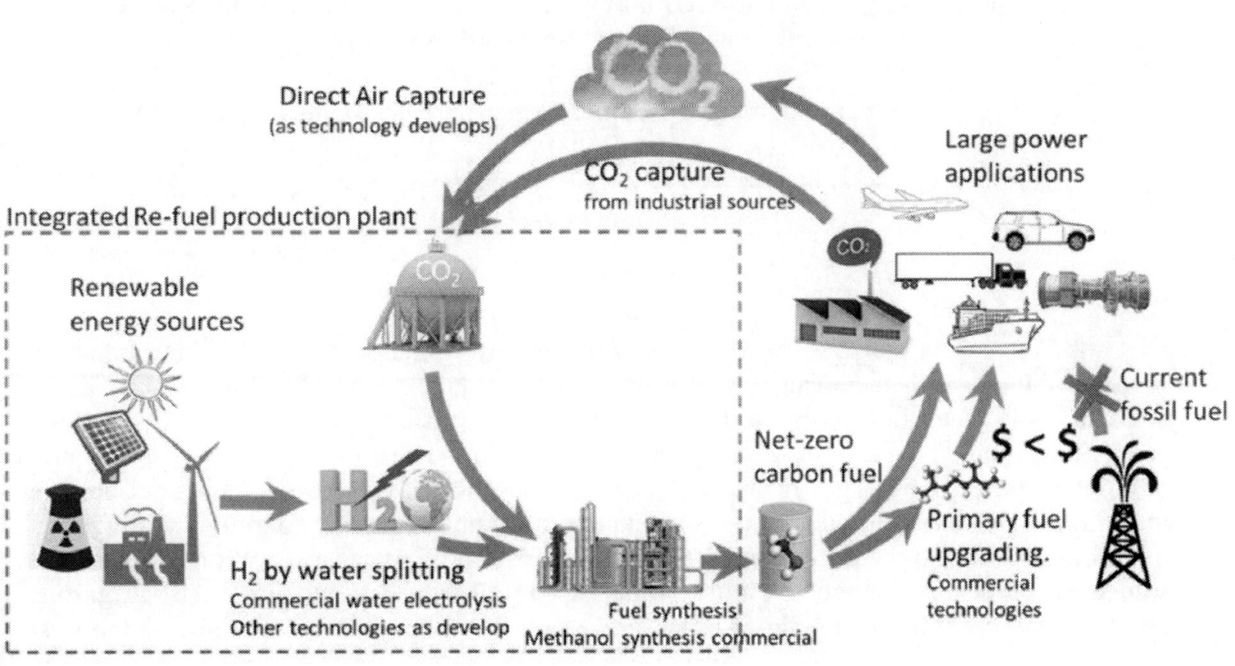

Direct Air Capture
(as technology develops)
CO_2 capture
from industrial sources
Large power
applications
Integrated Re-fuel production plant
Renewable
energy sources
Current
fossil fuel
Net-zero
carbon fuel
$ < $
Primary fuel
upgrading.
Commercial
technologies
H_2 by water splitting
Commercial water electrolysis
Other technologies as develop
Fuel synthesis
Methanol synthesis commercial

Multi-time energy integration and CO_2 management approach for synthetic natural gas production and storage using methanol as energy carrier

Meire Ellen Gorete Ribeiro Domingos[a], Daniel Flórez-Orrego[a], Dareen Dardor[a], François Maréchal[a]

[a]*École Polytechnique Fédérale de Lausanne, Valais, Sion, Switzerland*
meire.ribeirodomingos@epfl.ch

Abstract

Given the increasing concern about the environmental impact associated with the production of power, fuels, and chemicals, significant efforts are being made to explore sustainable alternatives to meet the energy needs. Biomass resources offer a promising solution for the energy transition, given their dynamic nature and the ability to be transformed into solid, liquid, or gaseous fuels. Biomass gasification, in particular, provides a versatile way to convert waste into energy. However, the adoption of such an approach must consider not only the mitigation of emissions but also the uncertainties associated with energy prices and carbon management strategies, especially in light of more stringent environmental regulations. Also, rigorous process synthesis and economic analysis are crucial for the decision making of the best pathways to boost bioeconomy.

On the other hand, methanol is a versatile chemical with significant potential in the energy sector. It can serve as a direct fuel or blended with gasoline and act as an intermediate molecule for various chemical syntheses, including hydrogen, methane, olefins, amines, acetic acid, dimethyl ether, and formaldehyde [1]. However, traditional industrial processes for methanol production rely on fossil sources, such as coal and natural gas, which have several drawbacks, including high energy and water consumption, pollutant emissions, nonrenewability, and future depletion [2]. To address these issues, alternative routes are being explored to decarbonize methanol's production supply chains, such as biomass thermochemical conversion via gasification [2]. Another advantage of methanol is that it can be stored as a liquid under ambient conditions, without the need for refrigeration or pressurization, which sets it apart from other energy carriers like hydrogen, liquefied natural gas, ammonia, and batteries [3].

This study proposes an integrated system for methanol production that combines different technologies and evaluates its potential as an energy carrier for synthetic natural gas (Figure 1). The production of methanol is achieved through the gasification of woody biomass, offering an alternative route to traditional methods. Additionally, the study considers the use of a water electrolysis system to produce hydrogen and generate syngas by combining it with the CO_2 stream from the syngas purification unit. The CO_2-H_2 mixture is sent to a reverse water gas shift reactor to adjust the composition before following to methanol synthesis and purification. The purified methanol can be stored and later converted to synthetic natural gas in a reformer, with the storage decision based on seasonal electricity and synthetic natural gas market costs.

The chemical processes synthesis, modeling and simulation are performed by using Aspen Plus® software [4]. The ultimate mass-based wood composition is set to 46.7%C, 6.02% H, 44.95% O, 0.17% N, 0.02% S and 2.14% ashes, whereas the mass-based proximate analysis is considered as 50% moisture (as-received), 14.32% fixed carbon, 83.54% volatiles, and ash in balance. The Battelle Columbus Laboratory indirect gasifier operating at atmospheric pressure and using steam as gasification medium is used for modeling the wood gasification system. This system avoids the dilution with nitrogen of the syngas produced, as the combustion and gasification processes occur in a separate double column system. After the produced syngas leaves the gasifier, a thermal catalytic cracking of the produced tar is performed. A fraction of

the char produced in the pyrolysis step is combusted to supply the heat required by the endothermic drying, pyrolysis and reduction reactions. After the produced syngas leaves the gasifier, it is cooled down and scrubbed with water, in order to remove the impurities that may affect the downstream equipment. The syngas composition must be adjusted considering the H_2:CO ratio equal 2. Thus, it is necessary autothermal reforming, shift and CO_2 capture processes in order to purify and upgrade the syngas composition.
In the methanol synthesis process, purified syngas is compressed to 90 bar and heated by the reactor outlet stream in a feed-effluent heat-exchanger. The syngas is then fed to an isothermal reactor operating at 90 bar and 210 °C. The reactor outlet stream is a gaseous mixture of methanol, water, and unconverted reactants which is cooled and flashed twice. The first flash cools the mixture to 30°C and 45 bar, while the second flash cools it to 3.5 bar, allowing for the separation of the condensable products and non-condensable reactants. The condensed stream proceeds to a distillation column operating at atmospheric pressure, producing methanol with purity above 99% wt.
The Peng-Robinson equation of state with Boston-Mathias modifications is mostly used as the thermodynamic model. The NRTL-RK property method (a nonrandom two liquid activity-coefficient model and the Redlich-Kwong equation of state for the vapor phase) is used to model the distillation section of methanol, which is suitable for describing the phases present in the system. The methanol synthesis is modeled using the LHHW (Langmuir-Hinshelwood Hougen-Watson) kinetic model reformulated for implementation in Aspen Plus [5].
Meanwhile, the OSMOSE Lua platform handles both the determination of minimum energy requirements and the solution of the energy integration problem. This optimization problem minimizes the resources consumption (wood, power) and, thus, the operating cost of the chemical plant, and also the investment required to purchase these new technologies (a.k.a capex), while satisfying some constraints, such as the heat balance at each temperature interval, the balance of consumed and produced power, the feasibility of the solution and the existence and size of the utilities units. In summary, it deals with the tradeoff of buying the technology and the operating revenues that may arise from its installation. To account for seasonal variations in energy prices, a multi-time integration approach combining different technologies and energy inputs is used to identify optimal operating conditions and arrangements that minimize energy consumption. The exergy method is used along with other financial indicators to determine whether and in which scenarios the proposed integrated setup would be more attractive.
As a result, the biomass to methanol production via gasification (see green section in Figure 1), presented a relative exergy efficiency that quantifies the deviation from the minimum theoretical exergy consumption necessary to make up the main industrial product, i.e. methanol, equal to 49.66%. Moreover, this system is producing 61.78 t_{CO2}/h from the CO_2 capture unit. The integration of technologies such as biomass gasification and electrolysis along with liquid fuel storage, may help tackling the intermittency of the renewable energy resources and increasing the economic revenues of the integrated plant. The optimal CO_2 management and methanol storage may ensure a reliable operation and synthetic natural gas availability even during the strained periods of the electricity grid and market conditions.

References

1. Leonzio G. State of art and perspectives about the production of methanol, dimethyl ether and syngas by carbon dioxide hydrogenation. Journal of CO2 Utilization. 2018;27:326–54.

2. Yang S, Li B, Zheng J, Kankala RK. Biomass-to-methanol by dual-stage entrained flow gasification: design and techno-economic analysis based on system modeling. Journal of cleaner production. 2018;205:364–74.

3. Irena, Methanol Institute. Innovation outlook: Renewable Methanol. International Renewable Energy Agency, Abu Dhabi [Internet]. 2021; Available from: https://www.methanol.org/wp-content/uploads/2020/04/IRENA_Innovation_Renewable_Methanol_2021.pdf

4. Aspentech. Aspen Plus V8.8. 2015;

5. Kiss AA, Pragt J, Vos H, Bargeman G, De Groot M. Novel efficient process for methanol synthesis by CO2 hydrogenation. Chemical engineering journal. 2016;284:260–9.

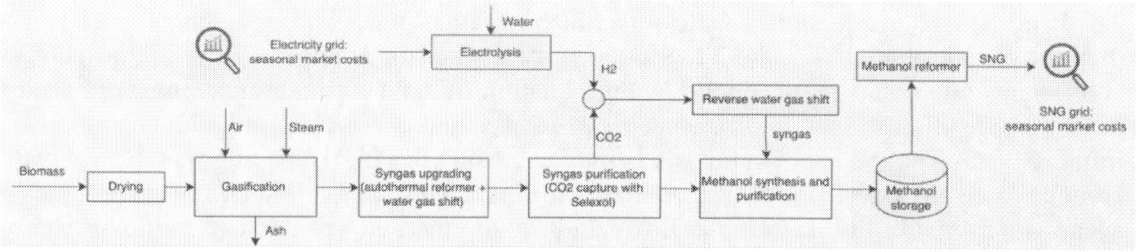

Figure 1. Concept for methanol production combining biomass gasification and electrolysis as an energy carrier to synthetic natural gas subject to different market conditions.

AUTHOR INDEX